INTELLECTUAL PROPERTY:
GUIDE FOR ENGINEERS

A Project for the
COMMITTEE ON ISSUES IDENTIFICATION,
AMERICAN SOCIETY OF MECHANICAL ENGINEERS

By the
COMMITTEE ON PUBLIC INFORMATION,
SECTION ON INTELLECTUAL PROPERTY LAW
AMERICAN BAR ASSOCIATION

Contents

FOREWORD

Intellectual property, the product of creative and hard working people, needs to be protected in ways that allow creative people to benefit from their efforts. Engineers create and use intellectual property and manage knowledge to their own benefit, as well as to the benefit of their company and society at large. To help engineers understand intellectual property issues, the Committee on Issues Identification of the Council on Public Affairs of ASME International collaborated with the Committee on Public Information of the Section of Intellectual Property Law of the American Bar Association (ABA) to create an intellectual property guide for engineers.

The guide is designed to provide a basic understanding of intellectual property issues, specifically those related to patents, trademarks, copyrights, and trade secrets. The guide is not meant to be a comprehensive statement of the law, and legal advice is always recommended when making a business decision.

As Chair of the Committee on Issues Identification, I would most like to thank the Honorable Gerald J. Mossinghoff, Senior Counsel, Oblon, Spivak, McClelland, Maier & Neustadt, P.C., and Chair of the ABA Committee, for his outstanding contribution to this guide and for his interest, inspiration, generosity, and patience throughout the project. He is the primary author and an expert in intellectual property issues.

I would also like to thank fellow members of the Committee on Issues Identification for their thoughtful contributions, particularly in reviewing and commenting on the drafts of the guide: Guy Arlotto, retired; George Flowers, Auburn University; William Hutzel, Purdue University; Dena Sue Potestio, National Conference of State Legislatures; Arnold Rothstein, Facilities Services Company; William Weiblen, retired; and Dave Wieland, John Deere Waterloo Works.

The contributions of the following people are also greatly appreciated: Sonya Engle, ASME Public Affairs Program Manager, for her efforts in coordinating the overall project; John Paul, Chair, ASME Technology & Society Division; Gloria C. Phares, Patterson, Belknap, Webb & Tyler; Phil Hamilton, Managing Director, Public Affairs; Chor Tan, Managing Director, Education; Harry Armen, immediate past Senior Vice President, Public Affairs; Yogi Goswami, Senior Vice President, Public Affairs; and Steven Lustig, past Leadership Development Initiative intern, Public Affairs.

Marc Goldsmith
Chair, Committee on Issues Identification

PART I

PATENTS

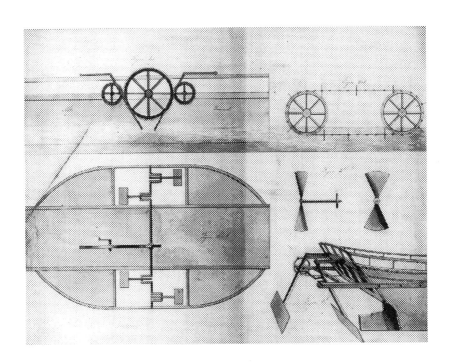

INTRODUCTION

In 1776, the philosopher/economist Adam Smith taught us that the wealth of any nation rested on three pillars: Labor, Capital and Natural Resources.[1] Our generation has added a fourth pillar — Intellectual Property in all of its forms. *Patents* protect new technology. *Copyrights* protect literary and artistic works, as well as computer software. *Trademarks* assure orderly commercial development and consumer protection. *Trade Secrets* provide competitive advantage to creative companies. Intellectual property provides important incentives in the burgeoning use of the Internet for e-commerce.[2] It is also an indispensable part of effective knowledge management — so critical in today's business and educational environment.

Because of their inherent role as creators and managers of new technology, engineers should have a basic understanding of the various forms of intellectual property and their underlying laws and governing principles. The purpose of this guide is to provide a summary of these forms of intellectual property and to point those seeking broader knowledge to the many sources of additional information, much of which is now on the Internet.

The United States patent system and copyright laws are as old as The Republic, having been established by the first U.S. Congress. Yet, each of these forms of intellectual property — together with the protection of trademarks and trade secrets — is directly and fundamentally involved in the accelerating pace of new developments in our new millennium both nationally and

---

[1]Adam Smith, *The Wealth of Nations: An Inquiry Into the Nature and Causes* (Modern Library 1994), 1776.

[2]Because of its importance and currency, a separate section of this discussion on *Intellectual Property and the Internet* is provided.

internationally. What should the scope of patent protection be for new software and business methods in e-commerce? Should transgenic plants and animals be protected by patents worldwide? Is patent protection for the human genome consistent with future advances in biotechnology? Can traditional trademark rights be accommodated in assigning Internet domain names? Can traditional copyright protection be enforced in an era of instantaneous and worldwide publication via the Internet? Will the new Federal Economic Espionage Act adequately protect privately owned trade secrets targeted by post-Cold War foreign espionage? Given U.S. leadership in the global economy, will the current national systems of intellectual property protection mature into effective multinational systems? Answers to these and similar questions are emerging in the recent developments outlined in this discussion. Engineers and business executives have a large stake in shaping the future policy and legal decisions, which — at the end of the day — must be tailored to serve them and their needs.

This discussion is focused under six headings: patents, copyrights, trademarks and service marks, trade secrets, intellectual property and the Internet, and international protection of intellectual property.

. . . the clear trend in modern times has been to increase patent protection . . . and make it more effective.

HISTORY

The United States patent system was established by the first U.S. Congress in 1790 under a specific grant of authority in the U.S. Constitution, specifically Article I, Section 8. That section states that "Congress shall have the power . . . to promote the progress of . . .useful arts, by securing for limited times to . . .inventors the exclusive right to their . . .discoveries." The U.S. Patent Office — now the U.S. Patent and Trademark Office or PTO — is one of the oldest of federal agencies, having been established in 1836 to provide an orderly and systematic examination of patent applications by professional examiners, all of whom are either engineers or scientists, and many of whom in addition have legal education and experience. The U.S. PTO is a key agency of the U.S. Department of Commerce employing more than 2500 examiners. Each of those, in turn, is assigned to examining patent applications in a very narrow area of technology. This organization permits each examiner to become an expert very quickly in his or her specific area of technology.

WHAT IS A PATENT?

A U.S. patent is a formal legal grant by the federal government that gives the inventor a legally enforceable right to exclude others from practicing the invention described and claimed in the patent. The federal government grants this right, for a term ending 20 years from the date of filing of an application for a patent, to encourage the public disclosure of technical advances and as an incentive for investing in research and development and commercializing the results. Thus, the overall progress of technical innovation is favored, while at the same time inventors are rewarded for their specific contributions. Like other forms of property, the rights symbolized by a patent can be inherited, sold, assigned, licensed or rented, mortgaged, and even taxed. Patents are national in character. Thus, a U.S. patent is enforceable only in the United States and its territories. If a company wants to practice an invention in foreign countries, separate applications must be filed in those countries or in regional patent offices that serve specific foreign countries. As pointed out later in the section on International Protection of Intellectual Property, the United States is a member of international treaties that facilitate the filing of patent applications in foreign countries.

What Conditions Must Be Met?

Congress has specified that a patent will be granted if the inventor files a timely application that adequately describes a new, useful, and unobvious invention of proper subject matter. The following conditions must be met.

☑ To be timely, an application must be filed within one year of certain acts, by the inventor or others, which place the invention in the hands of the public, i.e., patented or published anywhere in the world, on sale or in public use in this country.[1] This one-year grace period, however, is not available in most foreign countries. A U.S. inventor who wants to obtain corresponding foreign patents must first file an application in the U.S. before any disclosure, whether in written or oral form, of the invention to the public.

☑ The description of the invention in the application must be complete enough to enable others to practice the invention ("enablement"). Moreover, the application must describe the best manner ("best mode") known to the inventor of carrying out the invention.

---

[1] Recently, the Supreme Court, in a decision of particular relevance to engineers and high-technology executives, defined what is meant by the "on sale" bar to obtaining a patent. Essentially, the Court held that an invention is considered to be "on sale" if prior to the "critical date" (one year before the application for a patent is filed): it is embodied in a product that is the subject of a "commercial

☑ The described invention must be new, i.e., not invented first by another or identically known or used by others in this country or patented or published anywhere in the world before the actual invention date (not the application filing date).

☑ The invention also must be useful, i.e., serve some disclosed or generally known purpose.

☑ Finally, the invention must be unobvious, i.e., the differences between the invention and the prior public knowledge in its technical field (known as "prior art") must be such that a person having ordinary skill in this field would not have found the invention obvious at the time it was made.

The proper subject matter of a patent is very broad — in the words of the Congress and the U.S. Supreme Court, "anything under the sun that is made by man."[2] Specifically included is any product, process, apparatus, or composition, including living matter such as genetically engineered bacteria, plants, or animals. Special provisions also permit patents directed to certain distinct and new varieties of plants (Plant Patents) and new original and ornamental designs for articles of manufacture (Design Patents). A recent decision of the Court of Appeals for the Federal Circuit, the court that hears all appeals in patent-related cases, specifically held that business methods implemented by computers constituted patentable subject matter.[3]

---

offer for sale"; and it is "ready for patenting." That an invention is "ready for patenting" can be demonstrated in at least two ways: (1) by proof of reduction to practice before the critical date; or (2) by proof that before the critical date the inventor had prepared drawings or other descriptions of the invention that were sufficiently specific to enable a person skilled in the art to practice the invention. *Pfaff Electronics Co. v. Well Electronics, Inc.*, 119 S. Ct. 304, 311-312 (1998). Engineering firms that submit proposals for conceptual and developmental work must keep these criteria in mind in deciding when they must file a patent application on their work.

[2] *Diamond v. Chakrabarty*, 447 U.S. 303,309 (1980) (quoting S. Rep. No.1979): H.R. Rep. No. 1923, at 6 (1952).

[3] *State Street Bank & Trust Co. v. Signature Financial Group, Inc.*, 149 F.3d 1368, 1375-1377 (Fed Cir. 1998).

WHY OBTAIN A PATENT

Most inventors seek a patent to obtain the actual or potential commercial advantages that go along with the right to exclude others. Given the high cost of research and development, the opportunity to recoup these costs through commercial exploitation of the invention often is the primary justification for undertaking research in the first place. Patent rights can be commercially exploited in two basic ways:

- *directly*, by the inventor's practice of the invention to obtain an exclusive marketplace advantage (where the patented technology results in a better product or produces an old product less expensively); and/or
- *indirectly*, by receiving income from the sale or licensing of the patent.

THE NATURE OF PATENT RIGHTS

It is important to note that a patent (i.e., the right to exclude others) does not give the inventor the right to practice the invention. The inventor can practice his or her invention only if by so doing does not also practice the invention of an earlier unexpired patent. While only one patent can be granted on a particular invention, it is easy to see how more than one patent could be infringed by making a single product. For example, consider that A has a patent on a new type of door and B invents an improved door of this type with a special lock. B could not sell the improved locking door since A's patent broadly covers all doors of this type. On the other hand, A could not incorporate the improved lock in his basic door since B's patent covers this combination. In these circumstances both A and B can be free to practice the best technology (locking door) only if each grants a patent license to the other.

The indirect exploitation of a patent may be exclusive, e.g., by selling all rights in the patent or granting an exclusive license. Licenses can also be nonexclusive, allowing many parties, including the inventor, to practice the invention simultaneously.

A patent may also provide commercial advantages in addition to the potential for an exclusive market position or licensing income. A patent often lends business credibility to start up ventures and can open doors to both the technical assistance and financing necessary to bring a new product to market. An improvement patent may also provide the barter necessary to cross license any basic patents held by others that block the path to market.

How to Obtain a Patent

Patents are obtained through a complex administrative proceeding in the U.S. PTO. Since the legal rules that govern this proceeding are quite extensive and often complicated, it is strongly recommended that an inventor seek the assistance of an experienced patent attorney before beginning this process.[4]

Before actually applying to the PTO, there are several important preliminary steps that should be followed to prevent possible loss or damage to future patent rights.

- *Proper Record Keeping.* One of the most important of these preliminary steps is proper record keeping. Since United States patents are granted to the first inventor, it may become necessary to prove when, prior to the filing of an application, the invention was made. This is best accomplished by making a complete record of the invention from the first idea right up through development of commercial products. The invention record should clearly describe the invention with words and pictures (photographs, sketches, drawings, etc.) and should explain fully how it operates or is used. Each page of the invention record should be signed and dated in ink by the inventor. The record should also be reviewed as it is made by at least one other trustworthy person who is capable of

---

[4]Attorneys registered to practice before the PTO must have education in engineering or science and must pass an examination administered by the PTO to demonstrate that they have qualifications necessary to provide quality professional service to inventors.

understanding the invention, who should sign and date the record under the notation "read and understood by . . .."

- *Preliminary Evaluation of Patentability.* Another important preliminary step is the determination of whether the invention is likely to be considered patentable by the PTO, and if so, whether a patent that might be granted would be broad enough in its coverage to be worthwhile in a commercial sense. Such a preliminary evaluation of patentability should be made by a patent attorney, based in part on the prior patents and other materials located in a search of relevant records in the PTO. While the attorney's opinion that the invention should be patentable is not a guarantee that the patent will be granted, if he or she finds that the invention probably is not patentable or economically worthwhile, the considerable cost and effort of going forward with the process can be avoided.

- *Preparation of a Patent Application.* The next step in the process of obtaining a patent is the preparation of a patent application. A patent application is a complex legal document, which must fully describe the invention with words and, where appropriate, drawings, and which includes claims that define the legal boundaries of the invention. It is essential to the validity of the patent, and its ability to adequately protect the invention, that the invention be described and claimed completely and precisely. Accordingly, the inventor should tell the patent attorney everything about the invention, including what problems it solves and what difficulties were overcome to make it work. Particularly important is the duty to tell the attorney about prior patents or other prior inventions, of which the inventor is aware, so this information can be disclosed to the PTO. The patent application will also contain a Declaration and Power of Attorney form, which the inventor must sign indicating that he has read and understood the application and affirming that he is the first inventor. The application and a filing fee are then formally filed in the PTO.

Congress has recently authorized a new form of preliminary patent application known as a *Provisional Application,* which can be filed at a lower cost and without claims and certain other formalities. This provisional application is not examined, but must be replaced by a conventional application within one year. The benefit of this new low-cost form of application is that it secures a legal filing date in the PTO, but yet does not count in determining the expiration date of the patent, which is measured from the date of filing of the conventional application.

The filing of an application for a patent does not create any enforceable rights since the courts will only stop an infringer after the patent is granted. Nevertheless, marking a device "Patent Pending" or "Patent Applied For" may discourage potential infringers since it notifies them that they may have to stop production once the patent is granted. It is unlawful to use such a notice unless an application for a patent is actually pending in the U.S. PTO. After the patent has been issued, it is also good practice to mark the products sold under the patent with the patent number because it gives the inventor certain important additional legal rights.

In the PTO, the application undergoes a process called examination. After an initial processing stage (which may take six to nine months or more), a patent examiner will review the application and write a letter (called an Office Action) commenting on it. The first Office Action often is a refusal to grant the patent, and the applicant then has an opportunity to modify the application to overcome the examiner's objections. With the inventor's help, the patent attorney will reply in writing to the Office Action, usually making some changes and arguing that others are not necessary. Typically, at least two such exchanges between the patent examiner and attorney are necessary to resolve all the legal and technical issues. In general, it now takes an average of two years from filing to complete the examination process. Under a law enacted in November 1999, pending patent applications will be published

18 months after their effective filing date unless the inventor certifies that he or she will not file a corresponding application in foreign countries. Before the application is published, the application is kept secret, i.e., only government personnel and persons authorized by the inventor are permitted to examine the file. Publication of the application under the new law may entitle an inventor to provisional royalties from those who copy the invention.

When the examiner is satisfied that the application is in proper form and its claims are allowable, the applicant is notified that a patent will be granted upon payment of final government fees. In order to keep the patent in force until it expires, it also is necessary to pay progressively higher maintenance fees at intervals of 3, 7, and 11 years after the original grant.

THE EMPLOYED ENGINEER AS INVENTOR

As a general rule, an engineer/employee owns the patent rights to his or her inventions, with two important — some would say overriding — exceptions:

- An engineer/employee must assign patent rights to his or her employer if he or she was initially hired or later directed to solve a specific problem or to exercise inventive skill.
- An engineer/employee must assign patent rights to his or her employer if he or she signed an assignment contract.

Whether or not the first exception would apply in any case, companies commonly use assignment contracts as a condition for employment.[5] If reasonable, the courts will enforce such assignment contracts, and an employer may file a patent application on an employee's invention if the inventor refuses to do so. Whether an employed engineer's invention is covered by an assignment contract depends on whether the invention is related to the engineer's duties or field of activity rather than where or when it was made, i.e., at home or while on vacation.

---

[5]Executive Order 10096 controls the ownership rights of the federal government in inventions made by federal employees.

Patent attorneys can be quite specific in determining when an invention was made, in the legal sense. Normally, assignment contracts cover only inventions made during the course of employment or for a reasonable time after employment. When an engineer leaves a job, on-the-record exit interviews are often quite helpful to all concerned to define the respective rights of the engineer and the former employer.

ENFORCEMENT OF A PATENT

While the patent grant makes the information in the application available to the public, the inventor has the right to prevent others from making, using, selling, or importing into the United States what is claimed for as long as the patent remains in force. Patents are enforced in the United States against private parties by the filing of a civil action in the U.S. District Courts around the country.[6] Actions for infringement by the U.S. government must be filed in the U.S. Court of Federal Claims in Washington, D.C. All appeals in patent cases — whether from the District Courts, the ITC, the U.S. Court of Federal Claims, or the U.S. PTO — are heard and decided by the U.S. Court of Appeals for the Federal Circuit in Washington. In turn, appeals from that Court can be taken to the U.S. Supreme Court, but the U.S. Supreme Court decides relatively few patent cases.

Enforcing a patent in litigation can be expensive and time consuming. For corporations, it involves a significant but normal business expense; but for independent inventors, the costs of litigating a patent may require entering into a joint venture or otherwise seeking financial support.

In determining whether one's patent is being infringed, one would look to the normal and appropriate sources of corporate intelligence: trade shows, technical literature, advertising, sales and marketing information, etc. Once a potential infringement is

---

[6]Actions to prevent the importation of an alleged infringing product can be filed with the International Trade Commission (ITC) in Washington, D.C., which will handle the request to exclude the product on an expedited basis.

suspected, an inventor should immediately seek the advice of a knowledgeable patent attorney. Failure to approach the suspected infringer properly could result in the inventor being sued in what's called a Declaratory Judgment action to hold the patent non-infringed, invalid, or unenforceable. Once litigation is begun, the rules of federal discovery require that each side be thoroughly informed of all relevant information regarding the inventor's and the alleged infringer's records and actions.

TRENDS IN PATENTS

Although there is some inevitable opposition in some quarters, principally by those who profit by copying the creative works of others, the clear trend in modern times has been to increase patent protection, have it apply to new fields, and make it more effective.

- In 1980, the Supreme Court extended patent protection to living organisms in the famous *Chakrabarty* case.[7] This led to patents on biotechnology products, including transgenic plants and animals.[8]
- In 1982, Congress established the Court of Appeals for the Federal Circuit to hear all appeals in patent cases nationwide. This single step eliminated geography-dependent patent decisions. Prior to the establishment of the Federal Circuit, decisions in patent cases depended as much on where they were decided as their individual merits.
- In May of 1998, the Federal Circuit clarified that patent protection was available for software and for inventions including

---

[7]*Diamond v. Chakrabarty*, 447 U.S. 303 (1980).

[8]With the issue of granting valid patents on living organisms no longer in doubt, the focus of debate will now concentrate on the patentability of inventions related to the human genome. This issue does not involve prosthetic devices or other mechanical/electrical medical devices, whether regulated by the Food and Drug Administration or not. Such devices have always been patentable and there is every indication that they will continue to be patentable without challenge.

business methods.[9] This has led to a quantum jump in patent applications on e-commerce.

- Internationally, Japan, Europe, and the United States are undertaking concrete efforts to make it easier for inventors to protect their inventions beyond their own borders, as will be outlined later in the section on *International Protection of Intellectual Property.*

PATENT ISSUES

As the patent system moves into new areas — biotechnology, genomics, software, and business methods — some question arises whether the U.S. PTO is equal to the task of providing quality examinations. And each area has required the PTO to adopt new procedures. The enormous increase in inventions related to mapping the human genome, for example, required the PTO to establish a professional search capability — using the latest in on-line worldwide databases — to assist the examiners. The same may happen with respect to software and business methods.[10] What the PTO uncovers as "prior art" is always augmented by what patent applicants must disclose to the PTO. Each applicant has a clear obligation to disclose material prior art to the PTO; failure to do so in any case renders any resulting patent unenforceable.

In the words of Abraham Lincoln — himself a recipient of a U.S. patent — the patent system "added the fuel of interest to the fire of genius."[11] As the patent system moves to protect new areas of technology, it continues to serve that very real purpose — for engineers and their peers in other areas of technology.

---

[9] *State Street Bank and Trust Co.,* supra, note 6.

[10] On March 30, 2000, the Commissioner of Patents and Trademarks announced new procedures to examine patents on business-methods software. These will include a second level of examiner review together with the use of new databases and additional training and consultation.

[11] President Abraham Lincoln, Address to the Springfield, Illinois Library Association at Morris Library, Southern Illinois University–Carbondale (Winter 1859).

PART II

COPYRIGHTS

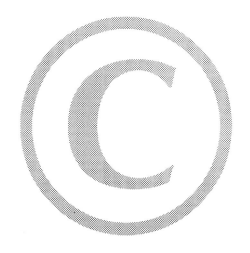

If copyright were enforced too strictly, it could obstruct goals that copyright was intended to protect — the creation of new works, education, and the growth of knowledge . . .

History

Like the U.S. patent system, copyright in the United States can trace its roots back to the U.S. Constitution, Article 1, Section 8, which states "The Congress shall have Power . . . To promote the Progress of Science . . . by securing for limited Times to Authors . . .the exclusive Right to their . . . Writings." Over time, the word "writings" has been interpreted to include a wide array of creative works. The purpose of the copyright clause in the Constitution and the laws that implement it is to encourage creative works by giving authors an economic incentive in the form of a copyright that permits the owner to exercise certain rights exclusively for a period of time. At the end of that time, the works enter the public domain for the use of all.

Although Congress passed the first federal copyright law in 1790, it was not until 1897 that the Copyright Office became a separate department of the Library of Congress, as it remains today. Since the first Federal copyright was registered in May 1790, more than 25 million copyrights have been registered. Works deposited with the Copyright Office make up a significant portion of the Library of Congress collection.

Until 1976, unlike copyright in much of the rest of the world, U.S. copyright law had required various "formalities" in order to secure copyright protection: publication with notice, registration, renewal of copyright registrations, and a requirement that a work be "manufactured" (printed) in the United States. In 1976, Congress passed the "new" Copyright Act, which finally dispensed with nearly all formalities, except the requirement that a work be published with notice. Even that requirement was finally abandoned in 1989, when the United States signed the major international copyright convention, the Berne Convention, which does not permit formalities as a condition of copyright.

WHAT DOES A COPYRIGHT PROTECT?

Copyright protects literary works, which the Copyright Act defines to include: musical works, including any accompanying words; dramatic works, including any accompanying music; pantomimes and choreographic works; pictorial, graphic, and sculptural works; motion pictures and other audiovisual works; sound recordings; architectural works; and computer programs.[1]

Copyright also protects compilations of material (even uncopyrightable material such as facts), but the protection extends only to the original selection or arrangement of the underlying material and does not give the author any rights in the underlying material. A sub-class of compilations is collective works, such as journals or magazines, where the underlying elements (articles, advertisements, editorial columns, and graphic materials) are themselves copyrightable.

Finally, copyright protects derivative works, which the Copyright Act defines as works based on one or more preexisting works, such as a translation, abridgment, condensation, musical arrangement, dramatization, or any other form in which a work may be recast, transformed, or adapted. For example, all sound recordings are derivative works of the musical compositions that they perform, and motion pictures are derivative works of their screenplays. However, the copyright in a derivative work extends only to the material contributed by the author of the derivative work and does not imply any exclusive right in the underlying material. The copyright in the derivative work is independent of the underlying work and does not affect or change the scope, duration, or ownership of copyright in the underlying work. As apractical matter, the owner of a copyright in a derivative work cannot exercise his or her rights without a license from the owner

---

[1]Copyright protects elements of a program that would be independently protectable, such as the source code and visual displays, but, as with any copyrightable work, copyright does not protect every element of a computer program. It may not, for example, protect functional interfaces that are commonplace, program logic, algorithms, or the concepts and ideas underlying the program.

of copyright in the underlying work, because exercising rights in the derivative work invariably involves exercising copyright rights in the underlying work as well.

Especially when dealing with non-fictional and scientific or technical works, it is important to keep in mind that copyright protects the expression of the author's ideas, but it does not protect the ideas or facts that are described or discussed in a work. The aim of copyright is to provide incentives to authors by protecting the author's own creative expression, but it does not give him or her a monopoly on the ideas or facts that are the building blocks of general knowledge. Copyright also does not protect methods, procedures, systems, inventions, methods of operation, regardless of the form in which they are described, explained or illustrated, or the mechanical or utilitarian aspects of a work of artistic craftsmanship.

HOW IS COPYRIGHT ACQUIRED?

The Copyright Act says, "copyright subsists," that is, arises automatically "in original works of authorship fixed in a tangible medium of expression." In order for a work to receive copyright protection, it must be "fixed" — on paper, videotape, disk, hard drive, audiotape, canvas, etc. — and it must be "an original work of authorship." That last phrase incorporates several concepts. First, the work must be creative (although the level of required creativity is very low); it must be original (not copied from another's work); and it must be the creation of a human author: materials produced solely by nature, plants, or animals, or by machines, are not copyrightable. An impromptu stand-up comedy routine is not protected because it is not fixed. Short words and phrases (including titles) are not copyrightable because they are not considered to have a sufficient level of creativity. Because U.S. copyright law no longer requires any formalities as a condition of acquiring a copyright, copyright arises automatically if a work meets the requirements of fixation, creativity, originality, and human authorship. Nothing more is required.

Who Owns the Copyright in a Work and How Is It Transferred?

In whom does that copyright arise? In the creator of the copyrightable work, who under the Copyright Act is referred to as the "author" — whether that creator is a writer, composer, or programmer. The creator of the work is the author and initial owner of the copyright from the moment of creation. If a work is a "joint work" — that is, prepared by two or more authors who intend that their work be merged into inseparable or interdependent parts of a unitary whole and intend to own the copyright together — then the authors are joint owners of the copyright in the work.

The only exception to that basic rule of ownership relates to works made for hire. A work is a work for hire under only two circumstances. First, it is made for hire if a regular employee working within the scope of his or her employment creates it, even if the work is not performed at the employer's workplace. In that case, the employee considers the employer the author and copyright owner of the work from the moment of creation and fixation. Second, a work is made for hire if and only if: (1) the work is one of nine categories of works;[2] and (2) the hiring party and the author agree that the work will be a work for hire and <u>both</u> sign a written agreement that the work is being commissioned as a work made for hire.[3] In that case, the hiring party is considered the author of the work from the moment of fixation. Unless the parties meet all

---

[2]The nine categories of works are: (1) a contribution to a collective work; (2) a part of a motion picture or other audiovisual work; (3) a translation; (4) a supplementary work ("a work prepared for publication as a secondary adjunct to a work by another author for the purpose of introducing, concluding, illustrating, explaining, revising, commenting upon, or assisting in the use of the work such as forewords, afterwords, pictorial illustrations, maps, charts, tables, editorial notes, musical arrangements, answer material for tests, bibliographies, appendixes, and indexes"); (5) a compilation; (6) an instructional text ("a literary, pictorial, or graphic work prepared for publication and with the purpose of use in systematic instructional activities"); (7) a test; (8) answer material for a test; and (9) an atlas.

[3]17 U.S.C. ß 101 ("work made for hire").

the statutory requirements, the work will not be considered work made for hire. If a work is not one of the ten statutory works, then it cannot be a work made for hire, even if the parties agree that it is. For example, a computer program is not one of the ten categories of works and cannot be commissioned as a work made for hire. If a work qualifies as one of the ten categories of works, but the parties do not both sign an agreement such conditions, then it will not be a work for hire, and copyright will arise in the author/creator, not in the hiring party.

Copyright is a personal property right that can be transferred and licensed. Ownership, transfer of ownership, licensing, and, in some cases, inheritance of a copyright, is governed primarily by the Copyright Act, although state contract law also applies. However, where there is a conflict between federal and state law — and sometimes where there is a conflict between federal law and the provisions of a contract or will — the provisions of the Copyright Act will preside.

Transfer of copyright ownership, which includes an assignment, exclusive license, mortgage, or other conveyance of the copyright (but not a nonexclusive license) is "not valid" unless it is in writing and signed by the owner of the rights conveyed. The Copyright Office does not prescribe any forms for transfers. The only requirement is that it be in writing and signed. An understanding between the parties or a confirming letter by the recipient of the rights is not sufficient. This writing requirement is very strictly enforced because it is meant to protect authors and other copyright owners from conveying rights orally or by course of business when they did not fully intend to do so. Similarly, any business that depends upon assignments or exclusive licenses of copyrights must ensure that those assignments and licenses are in writing; if not, they are not valid. The law provides for recordation in the Copyright Office of transfers of copyright ownership. Although recordation is not required to make a valid transfer between the parties, it provides certain legal advantages and may be required to validate the transfer against third parties.

The exclusive or nonexclusive grant of a transfer or license of copyright or of any of the rights of copyright in a work, even if assigned "forever" or "for the term of copyright" may not be forever or for the term of copyright. The law gives an author (and if an author is no longer living, certain specified family members) the right to terminate such a transfer or license 35 years after it was made. The result is that if the transferee/licensee wants to continue exploiting the rights it had, it must negotiate a new agreement with the copyright owner. This "termination right" is intended to give an author the opportunity to negotiate a new agreement and benefit from the value of his or her work that might not have been appreciated when the grant or license was originally made. This termination right cannot be waived in advance by contract. This termination right also does not apply to a work made for hire.

How Long Does a Copyright Last?

The term of copyright is complicated by the fact that it depends upon the date the work was created. Different rules apply to works made before January 1, 1978, the effective date of the current Copyright Act, and different rules apply to works made for hire. Generally, works created by an individual author (or authors, if jointly created) after January 1, 1978, are protected by copyright for 70 years past the death of the last surviving author. If a work was made for hire after that date, the term is 95 years from first publication or 120 years from the year of creation, whichever expires first.

Calculating the term of copyright for works created before 1978 under the "old" (1909) Copyright Act is more complicated. The terms for works created before 1978 and protected by federal statutory copyright on December 31, 1977, are extended for 95 years from the date of the original copyright (either publication with notice or registration of an unpublished work). For works not protected by federal statutory copyright (that is, unpublished works protected by state common law copyright), the term of copyright extends until December 31, 2047, if the work is published before December 31, 2002; otherwise the term expires at the end of 2002.

What Copyright Can Do for You

Copyright protection gives the author of a work the exclusive right to exercise (or license the right to exercise) the following rights in a work:

- reproduction of copies of a work;
- distribution of a work, which includes sale or other transfer of ownership, rental, lease, or lending;
- preparation of derivative works;
- public performance;
- public display.

Unless there is some statutory exception, no one can exercise any of the copyright rights in a work without a license from the copyright owner.

If copyright were enforced too strictly, it could obstruct goals that copyright was intended to protect — the creation of new works, education, and the growth of knowledge — or with constitutional rights such First Amendment rights. For example, if no one could reproduce any copyrightable work without an author's permission, a book reviewer might be prevented from reviewing the work in a meaningful way; a teacher might not be able to show his or her students copies of a noteworthy article that appeared in the newspaper on the morning before class; and a newspaper editorialist might be prevented from criticizing a political candidate's recent book. In order to balance these interests (and other unintended consequences of a too-strict enforcement of copyright) against the rights of copyright owners, the law provides a number of exceptions to the rights of a copyright owner. Some of these are very specific and deal with specific situations. The best known and more general exception is *Fair Use*: the right to make "fair use" of a work protected by copyright. Fair use limits the otherwise absolute copying and publication rights of a copyright holder by allowing the reproduction of a small portion of a work for "fair" purposes, such as comment and criticism, news reporting, teaching, scholarship, and research. Whether a particular use is "fair" cannot be easily predicted. It depends on the particular facts of the use: the purpose of the use, the nature of the underlying work, the amount used in proportion to the work as a whole, and effect on the potential market for the work are all factors that help determine whether or not the use is fair.

REGISTRATION AND COPYRIGHT NOTICE

As explained above, a work need not be registered in order to receive the protection of copyright laws. However, to encourage registration, the Copyright Law grants significant benefits to registration and conditions certain rights of enforcement on it. First, the owner of a copyright in a United States work cannot bring a lawsuit to enforce it against an infringer unless the work is registered. Although a work can be registered on the eve of a suit (at much greater expense), the law provides incentives for doing so earlier. Registration before or within five years of publication is prima-facie evidence of the validity of the copyright. If a work is registered before infringement occurs or within three months of first publication, then the copyright owner is entitled to statutory damages (which may be a helpful measure of damages where actual damages are difficult to prove) and entitles the copyright owner to recover attorney's fees from the infringer. This threat is often very effective in stopping an infringement, but it is not available if registration is not applied for until after the first infringement. If the copyright in a work is registered, the Customs Service is authorized to assist in preventing the importation of infringing copies of the work. For these reasons, registration of a work at the time of publications is recommended.

To register a copyright, an applicant is required to send the following, in one package, to the Copyright Office:

- a completed two-page application form that asks for basic information about the work, the author, and the copyright claimant (who may be the same);[4]

---

[4]Forms are available on the Copyright Office web site at http://lcweb.loc.gov/copyright. There are different application forms for different

- the appropriate filing fee ($30 through June 30, 2002);
- a deposit of two copies for a published work; one copy if unpublished.[5]

Once the application is processed, the copyright holder receives a certificate of registration, to which the Copyright Office has added a registration number and date and which is endorsed with the name and signature of the Register of Copyrights. Once approved, registration is effective as of the day the Copyright Office received the complete application.

A copyright notice is not necessary to obtain copyright protection, but placing a notice on one's copyrighted works is useful because it gives notice that the work is protected (as opposed to being in the public domain). It also can help defeat a defense of innocent or unknowing infringement, and may be necessary for copyright protection in some foreign countries. The three proper elements of a copyright notice are:

- the symbol ©, or the word "Copyright", or the abbreviation "Copyr.";
- the year of first publication of the work; and
- the name of the copyright owner of the work.

If you are circulating an unpublished work to a few people for comment or to consider publication, the copyright notice should not include the year; the year indicates the year of first publication and is inappropriate until the work is published to the general public.

---

classes of works. The Copyright Office intends that authors should be able to register works without the help of a lawyer, and its web site provides extensive explanatory material about how to register a work.

[5]Some types of works may require different submissions. For example, registration of a computer program requires deposit of portions of the source code. The Copyright Office provides special procedures for deposit of less than all the source code in recognition of the fact that deposit of source code could reveal trade secrets.

.

TRADEMARKS AND SERVICE MARKS

The best way to protect a company's investment in trademark or service mark rights is by obtaining a federal registration.

WHAT IS A TRADEMARK?

How does the field of trademark law impact the practice of an engineer? Any company that sells goods and services has to call its goods or services something so that consumers can distinguish the maker of a product or the provider of a service from others in the same field. Under the Federal Trademark Act, also known as the Lanham Act, a *trademark* is defined as

> any word, name, symbol or device or any combination thereof which is used by a person or which a person has a bona fide intention to use in commerce and applies to register on the principal register established by this Act, to identify and distinguish his or her goods, including a unique product, from those manufactured or sold by others and to indicate their source.[1]

This definition was derived largely from the common law definition developed by the courts many years before the enactment of the Lanham Act in 1946. The Lanham Act defines a *service mark* in essentially the same way except that a service mark is designed to identify and distinguish the *services* of one person from the *services* of others. Thus, a trademark or service mark could be a word such as EXXON or KODAK, a person's name such as TOMMY HILFIGER or GARTH BROOKS, a symbol such as the NIKE "Swoosh" or McDonald's "Golden Arches," or a device, such as the shape of a Coca Cola bottle, the roar of the MGM Lion or even distinctive visual movements of animation, such as the TRI-STAR Pictures flying horse at the beginning of a movie, or any combination of these things. It can even be the particular color of a product, such as pink fiberglass insulation. The

---

[1] 15 U.S.C. ß 1127.

Lanham Act also defines two other types of registrable marks, namely "certification" and "collective membership" marks. A *certification mark* is "any word, name, symbol or device or any combination thereof which is used by a person *other than the owner of the mark* . . . to certify the regional or other origin, material, mode of manufacture, quality, accuracy, or other characteristics of such a person's goods or services, or that work or labor on the goods or services was performed by members of a union or other organization." A *collective membership mark* refers to a trademark or service mark used by members of "a cooperative, an association, or other collective group or organization . . . and includes marks indicating membership in a union, an association or other organization." A common example of a certification mark would be words alone or words with designs certifying that an agricultural product, e.g., rice, watermelons, chickens, etc., are produced or grown in Arkansas. A *collective membership mark* is a mark that an organization might use to designate or identify members of the organization, e.g., the Rotary Club or a professional society such as the *American Society of Mechanical Engineers*.

Unlike patents, which are a grant from the federal government giving the inventor or patent holder the right to exclude others from making, using, or selling a patented device, trademarks and service marks are not created by or granted by the federal or any state government under the Lanham Act, but are recognized as soon as the owners of the mark uses the mark to identify goods or services for sale to the public. Therefore, federal or state registration of a trademark is not necessary in order for a company to own, use, or even enforce a trademark or service mark. However, federal registration confers benefits to the owner of the registration that are not otherwise available to the owners of unregistered "common law" marks.

Aside from the federal registration system through the U.S. Patent and Trademark Office, each state maintains a separate state registration system patterned essentially after the federal system. Most state trademark offices are administered by the Secretary of State's office or whichever state office handles the filing of corporation documents. State registrations are enforceable only by the courts in the particular state where issued. By contrast, a federal registration is enforceable nationwide. There is no prohibition against registering the same trademark or service mark federally as well as in any particular state.

Not all words, symbols, designs, slogans or combinations thereof used in the sale or advertising of goods or services function as a trademark or service mark, and it does not matter what the owner intended. For example, the Lanham Act does not permit the registration of trade names or promotional advertising campaigns for a business. Titles of single works, such as the title of a single book, movie, or record album, cannot be registered federally; however, the title of a series of works, i.e., two or more, may be registered.

Product configurations that are purely functional or utilitarian in nature cannot be registered or protected under trademark law. However, product configurations that are not purely utilitarian in nature, that may be ornamental or decorative in appearance, and that are either inherently distinctive or have acquired distinctiveness through long use or widespread advertising and promotion may be protected by trademark law if consumers recognize the product configuration being associated with a particular manufacturer. Think of the unique shape of a COCA COLA bottle, for example. While the bottle is utilitarian in the sense that it is designed to hold a certain amount of liquid, there are numerous ways competitors can make or shape a bottle to perform the same utilitarian function which does not have to look like a COCA COLA bottle. On the other hand, a device or product may have a unique design, but the design is so inherently utilitarian and necessary to the function of the device that no other manufacturer could make a competing product without using the same design or shape. For example, fiberglass insulation was once invented and patented. However, the natural color of the insulation as a result of the manufacturing process is generally a grayish-white. No one could claim the natural color or insulation as a trademark because that could essentially prevent others from making a competing product. However, if a dye is added during manufacture to make it pink or some other color, then the color could be protected under trademark law if it is distinctive, distinguishes the product as to its source in the minds of consumers, and provides no part in the essential utilitarian function of the insulation.[2] If a utility patent has ever been used for the particular device and the feature claimed as a trademark was claimed in the patent as being necessary to the utilitarian function of the device, it will not be protected as a trademark. This is to prevent utility patent holders from extending their exclusivity in a patented device longer than the term of the patent grant.

THE IMPORTANCE OF REGISTERING A MARK

As mentioned before, trademarks provide consumers with a method of distinguishing the goods or services of one company from another. If consumers like the quality of the product, they will generally want to buy it again. This favorable acceptance is an intangible benefit to the company called "good will." A company that spends a lot of time and money developing a strong identity with the consumer, e.g., business goodwill, should be very interested in keeping others from stealing or attempting to trade on the company's good will.

The best way to protect a company's investment in trademark or service mark rights is by obtaining a federal registration. The benefits of owning a mark registered on the *Principal Trademark Register* of the United States include providing the owner with nationwide priority rights from the filing date of the application for registration. A federal registration on the Principal Register also provides the owner with presumptive evidence of the validity of the registration, the registrant's ownership of the mark, as well as

---

[2] Indeed, the Supreme Court held in *Wal-Mart Stores, Inc. v. Samara Bros., Inc.*, 120 S. Ct. 1339, 1344-1346 (2000), that the configuration of a product itself (as opposed to its packaging) must have acquired distinctiveness through use in the marketplace in order to receive trademark protection.

the right to use the mark everywhere in the United States. A federal registrant can use the ® symbol to notify others that the mark is registered with the PTO. Companies that have applied, but not yet received, their federal trademark registration cannot use the ® symbol, nor can the owners of state trademark registrations or unregistered common law marks use it. In such cases the letters "TM" for trademark or "SM" for service mark can be used to notify others that such terms or logos are claimed to be trademarks or services marks as the case may be.

Federal registrants can also file a copy of the registration with the U.S. Customs Service to block the importation of products that infringe the owner's trademark or service mark rights. Criminal penalties can be imposed against infringers who counterfeit federally registered marks. Finally, a federal registration or an application for Federal registration can provide a company interested in selling goods or services abroad with the right to obtain registrations in foreign countries pursuant to international treaties in which the United States participates.

OBTAINING A FEDERAL REGISTRATION

To obtain a federal trademark registration one must file an application for registration with the PTO. For domestic applicants, registration can be based either on actual use in commerce or on a bona fide *intent-to-use* a mark in commerce. If the application is based on "use" in commerce, the applicant must include specimens such as labels or packaging for goods or advertisements for services showing how the mark is used in commerce.

The other basis for filing is based on the applicant's bona fide intent-to-use the mark in commerce. This filing basis was added to the Lanham Act in 1989 to give companies that had a bona fide intent-to-use a mark in commerce a right to file an application and obtain priority in a mark prior to actually using the mark in commerce. While an application based on "intent-to-use" will not be registered until the mark has actually been used in commerce by the owner, it is designed to notify third parties that those who adopt or use a similar mark after the filing of the intent-to-use application do so at the risk of being junior in priority in the right to use the mark. The PTO's processing of applications based on use or intent-to-use are essentially the same.

The Examination Process — Searching for Conflicts

After application for registration of a trademark or service mark is filed with the PTO, it is assigned to a Trademark Examiner who will search the mark for conflicts with marks previously registered on the Federal Principal and Supplemental Register as well as with prior pending federal applications. The PTO will not search or refuse registration to marks that are unregistered "common law" marks or marks that are registered with any state trademark office. The examiner will also review the application for compliance with other statutory and regulatory requirements. If a conflict is found with a registered mark, the examiner will issue an Office Action refusing registration based on likelihood of confusion under the Trademark Act. There are many factors the examiner, as well as the courts, considers in determining the likelihood of confusion between marks. Generally, the PTO will compare the mark for similarities in sound, appearance, and commercial impression or meaning. But it is not enough that the marks be similar in this regard. The marks must also be used on either the same or closely related goods or services, such that if consumers were to encounter the products or services in the marketplace they would think that the goods and services of the two companies originated from the same source. Therefore, if the goods or services of the companies are not related, the marks are not considered confusingly similar even though both companies may be using the same word or symbol as their respective marks. For example, would consumers think that DELTA airlines and DELTA water faucets originate from the same company? If not, then both companies can use and register the same word mark with the PTO.

On the other hand, if the goods and services of the two companies are closely related, then the degree of similarity between the marks need not be as great; and it would not be a defense to say that the marks are not exactly the same, e.g., using the word FOURD for automobiles or X-ON for gas stations. The likelihood of confusion can also be found where there are some differences in the nature of the goods and services. For example, clothing may be considered closely related to purses and jewelry because the same companies often make such goods and sold under the same marks and are often encountered by consumers in the same stores. Therefore, use of the same or similar mark by one company on a watch and by another on a sports jacket may be considered similar.

If a conflict is found with a prior pending federal application, the examiner cannot refuse to register the mark until the prior pending application actually is registered. The examiner will advise the applicant of the prior pending application and may suspend prosecution of the application until the pending application is either registered or abandoned. The period of suspension could last for several years.

TYPES OF REGISTRABLE AND NON-REGISTRABLE MARKS

The likelihood of confusion is only one possible ground for the PTO to refuse to register a trademark or service mark. Assuming there is no conflict with a Federally registered mark or pending application as discussed above, an applicant's mark still may not be registered for a variety of other reasons. Some of these reasons prevent marks from being registered under any circumstances while some will allow registration only if the mark has acquired distinctiveness, also referred to as "secondary meaning."

A *generic term* — i.e., a term that is the name of the class or genus of which the product or service is a member — can never be a trademark or service mark. Under this rule, if a valid trademark becomes generic in use, it loses its ability to serve as a trademark. Famous examples include "linoleum" for a floor covering or "escalator" for a moveable set of stairs. Also, marks that cannot be registered under any circumstances include scandalous and immoral marks such as obscene, offensive, or indecent words or depictions. Marks that are deceptive as to the material composition of the good or the geographic origin of the goods that would materially influence the purchasing decision cannot be registered. Examples of such marks might be a mark that uses the world "SILK" for clothing that does not contain silk or "Paris" for perfumes from Peoria. Since consumers are generally willing to pay a premium for clothing made from silk or perfumes from Paris, France, terms such as these would materially influence the purchasing decision for these particular products. Marks that disparage or falsely suggest a connection with persons, either living or dead, institutions, beliefs, or national symbols cannot be registered. For example, the mark BO BALL for a football with baseball-type lacing was once found to falsely suggest a connection with dual professional baseball/football player, Bo Jackson. Likewise, an application for MARGARITAVILLE was found to falsely suggest a connection with the singer Jimmy Buffet, who was famous for his recording of the song, "Margaritaville."

Marks that consist of the United States flag or coat of arms or any flag or coat of arms of any state or foreign country cannot be registered. Marks consisting of a name, portrait, or signature that identify any living individual cannot be registered without that individual's written consent and, in the case of presidents of the United States, neither during their lifetime nor during the lifetime of any surviving spouse without their consent. Therefore, while someone may be able to use and register GEORGE WASHINGTON as a mark for cherry pies, they would not be able to register GEORGE BUSH.

Marks that are geographically and deceptively misrepresent the origin of the goods or services — that is, marks that falsely represent the geographic origin of the goods but that *would not* materially influence the purchasing decision — can no longer be registered unless they were in use and had acquired distinctiveness prior to December 8, 1993. An example of this might be TENNESSEE TOMATOES for tomatoes grown in Arkansas. Most consumers will buy tomatoes from anywhere, particularly if they are on sale. The geographic deception, therefore, plays no part in the purchasing decision. However, the implementation of the North American Free Trade Agreement (NAFTA) on December 8, 1993, made these types of marks first used after that date just as unregistrable as deceptive geographic marks that do materially influence the purchaser.

Finally, generic names for good and services, such as SLIDE RULE for slide rules or MECHANICAL ENGINEERING SERVICE COMPANY for a company providing mechanical engineering services, cannot be registered at all for the exact goods or services they identify. However, just because a word sounds generic in one context does not mean it cannot be arbitrary or fanciful in another. For example, while SLIDE RULE may not be registrable for slide rules it could be registrable as the name of an alternative rock band. Likewise, MECHANICAL ENGINEERING SERVICE COMPANY could be registered as a mark for a pizza restaurant.

Marks that are not immediately registrable, but that can be registered after acquiring distinctiveness or secondary meaning, include marks that are merely descriptive of a feature, function, purpose, or characteristic of the goods or services, such as MECHANICAL ENGINEERING TIMES, for a publication about mechanical engineering subjects. Merely descriptive marks differ from generic marks in the sense that a generic mark identifies a specific genus or type of product or service while merely descriptive marks describe features or characteristics of the product or service. Other types of marks that require secondary meaning are those that geographically describe the origin of the goods or services, e.g., TENNESSEE TOMATOES for tomatoes grown in Tennessee, and marks that are primarily merely surnames, e.g., SMITH, JONES or LOPEZ. However, surnames that have a significant alternative meaning can be registered notwithstanding the fact that they are in fact surnames, e.g., LOOK, LAMB or PATRICK. The key is whether consumers primarily view the mark as a surname or as something else. Also included in this category are non-traditional trademarks such as product or packaging configurations, colors, sounds, movements, or scents, all of which are non-utilitarian (not purely functional features) and not inherently distinctive. All of these types of marks can be registered on the Principal Register if they have been used in commerce for a significant period of time, usually at least five years, and/or there has been a significant amount of money spent on advertising and promoting the products or services to consumers.

THE SUPPLEMENTAL REGISTER

If a mark is refused registration on the Principal Register because the mark is either merely descriptive, geographically descriptive, primarily merely a surname, or because it is a configuration or ornamental feature of the goods or packaging for the goods that is not purely utilitarian and is not inherently distinctive, and the mark has not yet "acquired distinctiveness," then such a mark may be registered on a second register called the Supplemental Register, which was set up specifically for these kinds of marks. However, the benefits of a registration on the Supplemental Register are not as great as with the Principal Register. For example, applications for marks that are based on "intent-to-use" cannot be made directly to the Supplemental Register. Registrations on the Supplemental Register are not considered prima facie evidence of the validity of the registered mark or the registrant's ownership or exclusive right to use the mark. A registration on the Supplemental Register cannot be used to block the importation of goods bearing infringing marks. However, the owner of a Supplemental Registration can use the federal registration symbol ® and it can be used as a basis for filing applications in foreign countries pursuant to international treaties. It can also entitle the owner to file an infringement lawsuit in federal court.

If during the initial examination of the application no conflicts are found with other federally registered marks or pending applications, and there are no other statutory refusals or informalities with PTO rules and regulations, the application will be approved for publication in the Official Gazette. Likewise, if an applicant files a response that resolves all outstanding issues in an Office Action, the application will be approved for publication. Once the mark is published in the Official Gazette, third parties have an opportunity to file an Opposition to registration of the mark. If no Opposition is filed, then the mark is either registered, if based on "use" in commerce, or a Notice of Allowance is issued if the application is based on "intent-to-use."

As with patents and copyrights, enforcing a federal registration is the sole responsibility of the owner of the registration. The PTO will not make any attempt to police the mark for the owner of the registration.

Term of Federal Registrations

Unlike patents and copyrights, trademarks can exist for as long as the manufacturer uses the mark in commerce, which could literally be forever. Once a federal registration is obtained, it is valid for an initial period of ten years; however, a registration may be renewed indefinitely as long as the mark is continuously used in commerce. After the fifth anniversary of the registration and before the sixth anniversary, an owner of a federal registration must file a declaration with the PTO that it is still using the mark and submit proof such as labels for goods or advertisements for services showing the mark is still in use in commerce. Anyone who is engaged in the manufacture and marketing of tangible goods or who performs services for others owns trademarks or service marks. Trademarks and service marks are valuable property rights that should be protected and guarded from theft or loss, the same as inventory or cash receipts. The most effective way for a company to protect its trademark and service mark rights is through federal registration of their trademarks and service marks with the PTO, supported by diligent monitoring of the marketplace for infringements. When infringements are found, the company should take immediate steps to enforce its trademark or service mark rights or risk the loss of those rights.

PROTECTING TRADEMARKS

If someone uses a mark or trade dress[3] that is confusingly similar to a mark or trade dress previously used by a prior party, the second comer can be found liable for trademark infringement, unfair competition, or both. If someone uses a mark or trade dress that dilutes a famous and distinctive mark or trade dress previously used by a prior party, the second comer can be found liable for dilution under the United States and state trademark laws. Lawsuits for trademark infringement or dilution are commonly brought in federal district courts, and the penalties for those found liable include injunctions prohibiting further complained-of conduct, monetary damages, and, in egregious cases, the imposition of the plaintiff's attorneys' fees.

---

[3]Although historically trade dress infringement consisted of copying a *product's packaging*, in its more modern sense it refers to copying the non-functional distinct appearance of the product itself, as well as its packaging.

PART IV

TRADE SECRETS

. . . in the absence of secrecy, the property disappears.

Originating in the early 19th century in the individual states of the United States, trade secret law protects secret business information against unauthorized use or disclosure by one who has obtained it through improper means or through a confidential relationship. State trade secret law is based on one of two distinct principles:

1. a property interest by the owner of valuable secret business information; and
2. an obligation of third parties to respect the confidentiality of valuable business information.

Although the Federal government obligated itself to respect privately owned trade secrets, trade secret law had been exclusively a matter for state courts prior to 1996. In October 1996, Congress passed the Federal Economic Espionage Act of 1996 (EEA), giving the Federal government the power to protect privately held trade secrets both through criminal laws and in civil actions by the U.S. Attorney General.

STATE TRADE SECRET LAW

In the interest of uniformity, most of the states have now adopted the *Uniform Trade Secrets Act* (USTA), drafted by the American Bar Association and recommended to the states. In 1974, the U.S. Supreme Court in a landmark decision[1] held that federal patent laws did not preempt the operation of state trade secret laws. This had been a matter of considerable confusion prior to the Supreme Court's ruling. The effect of the ruling is that all of the states now protect trade secrets.

---

[1]*Kewanee Oil Co. v. Bicron Corp.*, 416 U.S. 470 (1974).

The USTA defines "trade secret" as:

. . .information, including a formula, pattern, compilation, program, device, method, techniques or process, that:

(i) derives independent economic value, actual or potential, from not being generally known to, and not being readily ascertainable by proper means by, other persons who can obtain economic value from its disclosure or use, and

(ii) is the subject of efforts that are reasonable under the circumstances to maintain secrecy.

As one noted scholar has stated, "in the absence of secrecy, the property disappears."[2] To qualify as a trade secret, then, three conditions must be met: the information must be eligible for protection, be secret and have commercial value.

The protection afforded by the patent system, on the one hand, and trade secret law, on the other hand, is quite different; indeed, a patent and a trade secret on the same technology are mutually exclusive. In return for a *public* disclosure of a new, useful, and unobvious invention, the patent system grants a limited period of exclusivity to the inventor — 20 years from the filing of a patent application (see the section on Patents). Trade secret protection can last for an indefinite period of time, but does not offer protection against a third party's discovery and use of the technology by fair and honest means. As the U.S. Supreme Court has stated in the decision referred to earlier:

Trade secret law provides far weaker protection in many respects than the patent law.

• • •

Where patent law acts as a barrier, trade secret law functions relatively as a sieve.[3]

---

[2]Milgrim on Trade Secrets ß 1.03 (1998).

[3]*Kewanee*, 416 U.S. at 489.

Trade secret law provides protection only for that material the owner has made reasonable efforts to keep secret. A well known scholar has identified four sets of procedures that should be considered in protecting trade secrets.

1. Information-Directed Procedures
- identifying sensitive information
- notifying all those who may have access to it
- marking documents and other media with stamps or legends
- denying access to people who do not need the information
- securing computerized data with passwords
- putting physical notices on terminals and electronic notices on computer files and locked storage of recorded and printed data
- encoding secret information that is physically accessible
- placing signs and locks at entry points to entire facilities and to specific locations; destroying discarded records (e.g., computer print-outs)
- restricting access to copying machines
- permanently erasing data from old computers being sold or discarded
- dividing a process into steps and separating the various departments that work on the several steps
- using unnamed or coded ingredients
- destroying laboratory samples and trash on the premises
- using checkpoints, self-locking doors, alarms, closed circuit surveillance television, log-in procedures, watchmen, vaults, and shredding machines

2. Employee-Directed Procedures
- signing confidentiality and non-disclosure agreements
- giving notice of what information is particularly sensitive
- conducting entry and exit interviews
- restricting access to those employees who "need to know" specific information
- sending post employment letters

The employer should require employees to execute agreements, which may be part of an employment contract, containing non-disclosure and confidentiality provisions, provisions for assignment of discoveries to the employer, and covenants not to solicit fellow employees to leave and not to solicit customers of former employers.

3. Visitor-Directed Procedures
- accompanying visitors
- issuing admission badges
- denying access to particularly sensitive areas
- notifying employees that visitors are present
- requiring visitors to sign confidentiality agreements

4. Marketing and Public Disclosure Procedures. Disclosure procedures should include:
- delegating review of all material to people well informed about the company's valuable information
- restricting the level of detail and explanatory value of materials intended for general circulation, such as advertising, speeches and articles for publication
- limiting the degree of disclosure, and including confidentiality provisions in materials intended for limited circulation such as customer handbooks, repair manuals, specifications, bids and proposals.
- including deliberate typographical or other errors in confidentially disclosed materials to make it easier to trace the source of misused information[4]

---

[4]Donald S. Chisum et al., *Understanding Intellectual Property*, pp. 352 (Matthew Bender 1999). This work provides an excellent treatment of all of the major forms of intellectual property in a single volume.

ECONOMIC ESPIONAGE ACT [5]

With the enactment of the Economic Espionage Act of 1996 (EEA), the federal government moved aggressively into the protection of private property rights in trade secrets. The overriding reasons behind the enactment of the legislation were the fully documented efforts of foreign governments to gain access to the trade secrets of U.S. companies in order to advance the economic interests of their private sector. However, the EEA is not limited to enforcement against foreign governments, or even foreign-based companies. It applies as well to any typical trade secret dispute involving purely domestic concerns. With fines up to $5,000,000 and imprisonment of up to 10 years for the domestic theft of trade secrets, the EEA creates a new and very powerful intellectual property regime that extends well beyond the U.S. borders.

Probably the most troublesome feature of the EEA relates to its applicability in instances in which an employee of company X, having knowledge of company X trade secret information, changes jobs to work for competitor company Y. When the employee performs work for company Y using skill and knowledge obtained during employment at company X, is the employee in violation of the EEA? How can company X protect itself from loss of its trade secrets? Conversely, how can the employee be expected to forget what he has learned when going to work for company Y? The problem is that the employee cannot simply forget the trade secrets of company X and must therefore attempt to compartmentalize the various bits of knowledge and expertise gained while in the employ of company X.

Early versions of the EEA included a passage stating that "knowledge, experience, training or skill that a person lawfully acquires during their work as an employee or independent con-

---

[5] See Mossinghoff et al., *The Economic Espionage Act: A New Federal Regime of Trade Secret Protection,* 79 Journal of the Patent & Trademark Society 191 (1997).

tractor" for another person was not included in the term *proprietary economic information*. This provision was later removed and the term changed to *trade secret*. The legislative history indicates that the EEA is not intended to be used to prosecute persons who use generic business knowledge to compete with former employers. Unfortunately, this still leaves an employee in a dilemma as to what among his business/technical knowledge is considered "general knowledge" and what is over the line and classified as a "trade secret." Further it leaves company Y in the difficult position of hiring the employee for his knowledge and expertise and then having to tell him that he can only use portions of it, thus potentially limiting his effectiveness on the job, or limiting his job mobility.

There have been fewer than a dozen cases brought by the Department of Justice under the Economic Espionage Act — all of which have resulted in convictions or are still pending — so many questions still remain regarding its scope and meaning. Nevertheless, the EEA is an important recognition by Congress of the importance of trade secrets to U.S. businesses competing in a high-technology global economy.

PART V

INTELLECTUAL PROPERTY AND THE INTERNET

Much of the burgeoning e-commerce on the Internet is directly and fundamentally tied to intellectual property protection.

PATENTS

As discussed earlier, the Federal Circuit, in State Street Bank, clarified that patent protection was available for software and business methods. Prior to that decision, various prior computer and business methods were held to be ineligible for patent protection based on the U.S. patent law (i.e., the inventions were "non-statutory").[1] However, the dividing line between those inventions that are the proper subject matter of patents and those that are not was left unanswered by *State Street Bank*.

Less than a year later, the decision of the Federal Circuit in *AT&T Corp v. Excel Communications*[2] was perceived by many to signal that the courts were liberalizing their interpretations of the types of inventions that are statutory, i.e., that can be protected under the patent statutes. In that case, the Federal Circuit held that, unlike in previous decisions, a physical transformation of a signal from one form to another is not required. With the transformation requirement removed, the PTO has seen a substantial increase in the number of patent applications. In fact, the Commissioner of Patents and Trademarks stated that since the *State Street Bank* decision, e-commerce related patent filings have increased 700%.

That increase also is consistent with the realization by Internet companies that being "first-to-market" is no longer a guarantee of success. Internet retailers, such as Amazon.com and PriceLine.com, have begun to obtain patents in an attempt to protect market share

---

[1]For a thorough discussion of the history of "software patents" see Gregory J. Maier et al, *State Street Bank in the Context of the Software Patent Saga*, 8 George Mason Law Review 307 (1999).

[2]172 F.3d 1352 (Fed. Cir. 1999).

against their rivals. On December 1, 1999, Amazon.com obtained a preliminary injunction[3] against BarnesandNoble.com for Amazon's one-click ordering technology. That injunction required that BarnesandNoble.com stop using the Express Lane feature on the BarnesandNoble.com website.

In addition to judicial opinions interpreting what types of inventions are statutory, Congress itself indicated a desire to have business methods be considered statutory in the American Inventors Protection Act of 1999. According to that Act, the U.S. patent law was amended to provide prior user rights (i.e., the right to continue using a patented process that the non-inventor used before the filing of the inventor's patent application), but only as applied to methods of conducting and doing business. As a result, it is clear that methods of doing business must be statutory; otherwise there would be no need for the exemption. The scope of the protection provided by the new "prior user rights" and its business implications are far from clear and will remain so until we have judicial interpretations of the new legislation. Meanwhile, despite this uncertainty, most companies engaged in e-commerce are pursuing patent protection systematically and aggressively.

COPYRIGHTS

The Digital Millennium Copyright Act of 1998

It is important to remember that copyrightable material available via the World Wide Web or otherwise accessible via the Internet or other interactive computer systems is protected by copyright. Although it is particularly easy to copy works that are available electronically, the full force of the copyright law still applies. The Digital Millennium Copyright Act (DMCA)[4] contains a number of provisions designed to better protect the

---

[3] *Amazon.com, Inc. v. BarnesandNoble.com,Inc.*, 53 U.S. Patent Quarterly 2d 1115 (D. Wash 1999).

[4] The Act added 17 U.S.C. §§ 117 (c), 512, and 1201-05, among others.

rights of copyright holders in the digital age. In particular, the DMCA does the following:

- Prohibits circumventing a technological measure that controls access to a work protected by copyright. This provision will become effective on October 28, 2000.
- Prohibits the making or selling of devices designed (or marketed) to circumvent technology intended to protect copyright works by preventing access to, or the copyright of the works. (Manufacturers are not required to address incidental features of products designed and marketed for other purposes.) There are a number of exceptions to these two prohibitions, including an exception for encryption and security testing; law enforcement; personal privacy concerns; and reverse engineering otherwise permitted by the copyright law.
- Creates a limited liability for online server providers (OSP, e.g., America On Line) for copyright infringement that takes place over their systems. This section specifically limits OSP liability for transitory communications initiated by someone other than the provider; routine systems caching; information available on websites, or other accessible "space," hosted by a provider, so long as the provider does not have knowledge of the infringement.
- Permits the copying of a legally purchased computer program if necessary for the repair and maintenance of the computer upon which it is installed. The copy of the program must then be destroyed once it is no longer needed.

TRADEMARKS AND INTERNET DOMAIN NAMES

As pointed out earlier in this discussion, a trademark or service mark serves to distinguish the maker of a product or provider of a service from those goods produced or services rendered by others. In the brick-and-mortar world, the law governing the protection of trademarks and service marks was relatively settled. A second comer who uses a mark that is likely to cause confusion with the pre-existing mark of another, or which would cause dilution of a famous mark of another, subjects himself to liability.

With the burgeoning digital world of the Internet, particularly the World Wide Web, came a new set of problems. As the computer network, which today is known as the Internet, began to take shape, number strings called Internet Protocol (IP) addresses were the vehicles by which remote users could find one another. The IP addresses are formed in strings of numbers grouped in threes and twos, such as 012.34.567.89. When the commercial world was introduced to the Internet, users found IP addresses to be cumbersome and nearly impossible to memorize.

Engrafted over the IP system, then, were a set of "name-handles" by which remote Internet users could conveniently locate and easily remember where to find one another. The name-handles came to be called "domain names" and the system for managing them was appropriately called the "domain name system" (DNS).

No sooner was the DNS established than devious individuals purchased, for relatively inexpensive prices, domain names that incorporated distinctive or well-known marks or the names of famous people. These individuals would then brazenly offer to sell these domain names to the true mark owners or the famous persons for exorbitant asking prices. If the mark owner or famous person refused to make the purchase, the punishment often was the use of the domain name in connection with pornography or directly and unfairly competing businesses. Existing trademark, dilution, and privacy laws were found lacking in their ability to effectively deal with these individuals, who came to be known as cybersquatters.

On November 29, 1999, the President signed into law the Anti-Cybersquatting Consumer Protection Act (ACPA). The ACPA added Section 43(d) to the U.S. Trademark Act and a new regime designed to protect distinctive or famous marks and well-known persons from acts of cybersquatting. The trafficking in domain names, registered in bad faith, is now illegal. The statute also has a separate provision that provides for lawsuits directly against the domain names, if their owners cannot be found within the United States. Moreover, the ACPA provides for a schedule of statutory damages against cybersquatters who improperly registered their ill-gotten domain names after the ACPA became effective.

In addition to the anti-cybersquatting laws enacted within the United States, there is an international procedural mechanism by which an aggrieved party can seek the transfer of a domain name that incorporates someone else's trademark or name. Originally, DNS management of generic top level domain names (gTLDs) was the exclusive province of one company, Network Solutions, Inc. (NSI) of Herndon, Virginia. NSI's informal system of resolving domain name disputes was found to be unworkable, and frustrated many companies who were the owners of distinctive or well known marks.

The business community complained loud and long to the United States government. It became readily apparent that management of the DNS and the resolution of domain name disputes could no longer be the province of one company. A non-profit California corporation was established, the Internet Corporation for Assigned Names and Numbers (ICANN), to which was transferred control over the DNS for gTLDs (for now, the domains ending with the suffixes .com, .net, and .org), as well as the establishment of a regime for resolving cybersquatting-type domain name disputes. This regime is now known as the Uniform Dispute Resolution Procedure (UDRP). ICANN may be found on the Internet at the URL (Uniform Resource Locator) *http://www.icann.org.*

ICANN has delegated responsibility for the operation of the UDRP to four entities, the World Intellectual Property Organization (WIPO) located in Geneva, Switzerland, the National Arbitration Foundation (NAF) located in Minneapolis, Minnesota, eResolution located in Montreal, Canada, and the CPR Institute for Dispute Resolution in New York City. A listing of relevant sites is provided below:

> http://arbiter.wipo.int/domains/index.html: WIPO's arbitration center

http://www.arb-forum.com: NAF
http://www.eresolution.ca: eResolution
http://www.cpradr.org: CPR Institute
http://www.icann.org/udrp/approved-providers.htm: listing of ICANN providers

ELECTRONIC DATABASES

With the advent of the Internet came another benefit, and its accompanying problem — the ability to transfer massive amounts of data from one remote computer to another with a key stroke or the click of a mouse. Institutions that have spent millions of dollars and countless person-hours collecting and compiling valuable databases have found that nefarious operators are able to instantly misappropriate these assets over the Internet at will.

The European Union already has issued a directive for the protection of databases. The selection and arrangement of the database is protected under copyright as a compilation. The contents of the database are protected by a generic "extraction right" for a limited period of years.

Under the Supreme Court's 1991 decision in *Feist Publications, Inc. v. Rural Telephone Service Co.*,[5] United States copyright law gives a very narrow scope of protection to databases. Under this decision, raw facts and individual pieces of data collected from publicly available sources are not by themselves protectable under United States copyright law, no matter how difficult or expensive it was to collect this information. Only the unique selection and arrangement of this data is protected as a compilation. Moreover, the Supreme Court said that its conclusion was required by the Constitution. The copyright clause in the Constitution allows protection only for original works of authorship, and individual facts and ideas are not works of authorship.

---

[5] 499 U.S. 340 (1991).

There are efforts within the United States Congress to give separate statutory protection to database collections. The results of these Congressional efforts remain to be seen, particularly when college/university, library, and other fair use advocates are lobbying heavily against the promulgation of such laws without sufficient countervailing protections. Engineers, of course, have a keen interest in the outcome of these deliberations. In many of their roles, they are the creators of databases and, thus, would want to see the "sweat of their brow" protected. On the other hand, they necessarily rely upon databases in their day-to-day professional work and would share the view that absolute protection may not be appropriate.

INTERNATIONAL PROTECTION OF INTELLECTUAL PROPERTY

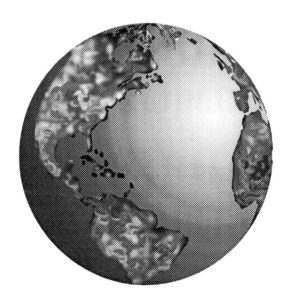

. . . of a new invention or a copyrighted work can travel around the world literally at the speed of light.

By its very nature, intellectual property is international in character. In today's world, the disclosure of a new invention or a copyrighted work can travel around the world literally at the speed of light. The improper use of a trademark on the Internet can mislead anyone having access to the World Wide Web. Thus, there is a strong international dimension to each of the major forms of intellectual property, but more needs to be accomplished in our global marketplace. A true leader in advancing international protection is the World Intellectual Property Organization (WIPO), a specialized agency of the United Nations headquartered in Geneva, Switzerland. The growth in the number of nations that belong to the WIPO is shown in Figure 1.[1]

PATENTS

The precursor of all modern-day multinational protection for intellectual property is the Paris Convention. The treaty was drafted in 1880 and became effective in 1884. The Paris Convention established the fundamental principles of "national treatment," "right of priority," and "special agreements," that have been incorporated in all subsequent multinational patent agreements. The increase in the number of nations adhering to the Paris Convention is shown in Figure 2.

The principle of "national treatment" requires member states to accord nationals of other member states the same advantages and rights under their domestic patent laws they accord to their own.

---

[1]For a discussion of the important work of the WIPO, see Mossinghoff and Oman, *The World Intellectual Property Organization: A United Nations Success Story,* 159 World Affairs 104 (Fall 1997).

WIPO Convention

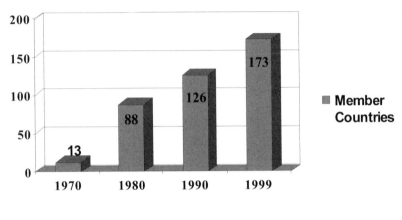

FIGURE 1

Paris Convention

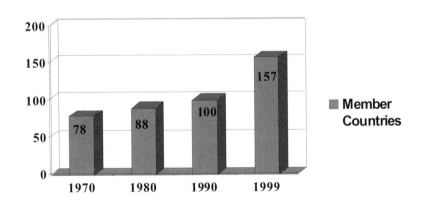

FIGURE 2

However, nationals seeking patent protection in a member country must comply with the domestic laws of the member country from which they seek patent protection. To augment the Paris Convention, in the late 1960s, the United States took the lead role in creating a new multilateral patent treaty to minimize duplicate patent applications and examinations worldwide. This effort resulted in the Washington Treaty of 1970, commonly referred to as the Patent Cooperation Treaty or PCT. This treaty constituted an important step toward rationalizing the filing of patent applications worldwide. The PCT entered into force on January 24, 1978, with 18 contracting states. One hundred and six states have now signed or acceded to this treaty (see **Figure 3**), which operates under the Paris Convention and is administered by the WIPO.

Internationally, efforts to harmonize patent laws are being addressed through a new *Patent Law Treaty*. Adopted at a diplomatic conference on June 1, 2000, the WIPO member states formally adopted this treaty to harmonize procedures for filing national and regional patent applications and maintaining patents. The treaty, which countries can now join, sets forth global rules for patent offices on such issues as obtaining a filing date for patent applications, electronic filing, representation of patent applicants, conditions for extension of time limits, and priority claims. There are now five regional patent offices, by far the most important being the European Patent Office, serving 19 European countries.[2] There is also movement to establish a global or world patent system, but we are only in the very early stages of actual development of such a system.[3]

---

[2]The other regional offices include two African offices, a Gulf Cooperation Council office in Saudi Arabia, and a Eurasian office in Moscow.

[3]See, e.g., Mossinghoff et al, *The World Patent System: Circa 20XX* A.D., 38 Idea 529 (Franklin Pierce Law Center 1998).

PCT

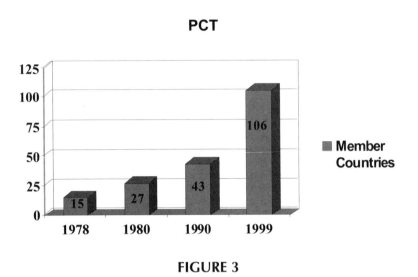

FIGURE 3

COPYRIGHTS

There is no such thing as an "international copyright." However, the United States is a party to a number of copyright agreements with other countries. The most comprehensive of these agreements are the *Berne Convention for the Protection of Literary and Artistic Works and the Universal Copyright Convention* (UCC), with approximately 142 and 97 member nations, respectively. Protection under each of these agreements is generally conferred if the work is published in, or the author is a national or domiciliary, of a member nation.[4] Under Berne, each member state follows the principle of "national treatment" — that is, the works of a foreign national are protected to the same extent the works of a

[4]Note that the UCC requires that a work bear a copyright notice in order to be protected.

Berne Convention

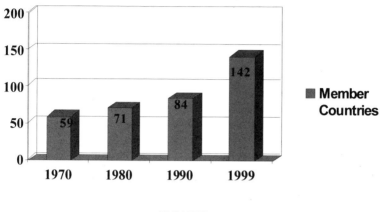

FIGURE 4

citizen of that member state are protected. More than 150 nations have some form of copyright agreement with the United States.[5] Figure 4 shows the growth in membership of the Berne Convention.

TRADEMARKS

Like patents and copyrights, the protection of trademarks and service marks is of national scope. Generally, owners must secure protection in each nation of interest. The *Community Trade Mark* (CTM) offers the opportunity to protect a trademark in all the

---

[5]Information regarding specific international policies and relationships can be found on the Copyright Office web site at http://lcweb.loc.gov/copyright/circs/. Of particular interest are Circular 38a, "International Copyright Relations of the United States," and the International Copyright Factsheet.

countries of the European Union (EU)[6] through the filing of a single application. Applicants may choose to file for a CTM at the Office for Harmonization in the Internal Market (OHIM) in Alicante, Spain, at the Trademark Office of any of the member states, or at the Benelux Office.

The most attractive feature of CTM registration is that it offers trademark protection in all 15 countries of the EU at a cost that is much lower than that of filing separate applications in each member state. Furthermore, use of the trademark is not required to secure registration or renewal. Also, use in a single member state is sufficient to maintain the validity of the CTM registration throughout the EU, and avoids its vulnerability to cancellation through non-use over a five-year period. Finally, a CTM application that is refused registration may be converted into national applications maintaining the priority of the original CTM application.

The *Madrid Protocol* (Protocol), in force since 1996, was adopted to facilitate the filing and international registration of marks in several countries.[7] A dispute regarding voting rights of member states between the European Union and the United States has now been resolved, and it is expected that the U.S. will join the Madrid Protocol next year. There are now 43 member countries of the Madrid Protocol, as shown in Figure 5. Under the Protocol:

- The applicant may base his application for international registration on a pending national application.

---

[6]The countries covered are Austria, Benelux (Belgium, Netherlands and Luxembourg), Denmark, Finland, France, Germany, Greece, Ireland, Italy, Portugal, Spain, Sweden, and the United Kingdom.

[7]The following countries are currently members of the Madrid Protocol: Austria, Benelux (Belgium, Netherlands and Luxembourg), China (People's Republic), Cuba, Czech Republic, Democratic People's Republic of Korea, Denmark, Estonia, Finland, France, Georgia, Germany, Hungary, Iceland, Kenya, Lesotho, Liechtenstein, Lithuania, Monaco, Mozambique, Norway, Poland, Portugal, Republic of Moldova, Romania, Russian Federation, Slovakia, Slovenia, Spain, Swaziland, Sweden, Switzerland, Turkey, United Kingdom, and Yugoslavia.

out/design of integrated circuits, trade secrets, and geographical indication (such as those used on wines). However, it does require adherence to various international intellectual property treaties, especially the Paris and Berne Conventions. TRIPS was a major global step forward. Next, we will move beyond international norms for national systems of protection to truly multinational systems themselves.

Madrid Protocol (Marks)

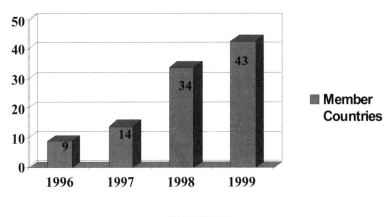

FIGURE 5

- Each national trademark office has 18 months to notify the World Intellectual Property Organization of objections to the international registration.
- In the case of a rejection or cancellation of a national application, an international registration may be transformed into national applications, benefiting from the original filing date or priority date, within three months from the date of cancellation.
- Applications may be in either French or English, whereas the Madrid Agreement, an earlier international trademark treaty, permits only French.

It is also envisaged that the Office for Harmonization in the Internal Market will become a party to the Protocol, thus enabling international registrations to be based on Community Trade Mark applications, or to receive European Community protection through an international registration.

TRADE-RELATED ASPECTS OF INTELLECTUAL PROPERTY RIGHTS (TRIPS)

The increasing importance of the relationship between intellectual property protection and international trade was recognized in the Uruguay Round of the General Agreement on Tariffs and Trade (GATT). During the Uruguay Round, negotiators recognized the benefit of intellectual property protection for both developed and developing countries. For developing countries, weak intellectual property protection discourages the necessary domestic investment in research and development that fuels economic development. Moreover, intellectual property protection no longer just promotes technological advancement domestically, it also provides means for nations to compete effectively in the global economy, spurring much needed technology transfer to developing countries.

In the Uruguay Round, member states ultimately agreed to an entire package of agreements: the creation of the World Trade Organization (WTO), amendment of the Dispute-Settlement Procedures, agreements on Trade in Goods and Agriculture, application of Sanitary Measures, agreements on Trade-Related Investment Measures and Countervailing Duties, Antidumping Measures, and Inspections and Customs Procedures. They also agreed to the Trade Related Aspects of Intellectual Property (TRIPS) Agreement.[8]

Under the TRIPS Agreement, comprehensive and binding obligations are now in place in every area of intellectual property protection: patents, copyrights, trademarks, industrial designs, lay-

---

[8]For an insightful analysis of the TRIPs negotiations, see Michael P. Ryan, Knowledge Diplomacy, Global Competition and the Politics of Intellectual Property (1998). See also Daniel Gervois, the TRIPs Agreement Drafting History and Analysis (1998).

MAJOR INTELLECTUAL PROPERTY INTERNET SITES

American Bar Association, Section of Intellectual Property Law
http://www.abanet.org/intelprop/

Intellectual Property Owners
http://www.ipo.org

U.S. Patent and Trademark Office (USPTO)
http://www.uspto.gov

U.S. Copyright Office
http://www.lcweb.loc.gov
Copyright Society of the U.S.A.
http://www.csusa.org

International Trademark Association
http://www.inta.org

American Intellectual Property Law Association
http://www.aipla.org

European Patent Office
http://www.epo.co.at/epo

Japanese Patent Office (JPO)
http://www.jpo-miti.go.jp

World Intellectual Property Organization (WIPO)
http://www.wipo.int

Conclusion

Engineers are directly involved with intellectual property in their day-to-day work. They are the creators of new technology, often working on research and development projects undertaken with the promise of exclusive rights in the results. And, they are often the users of new technology created by others. As they move to senior management positions, intellectual property can drive their most important corporate decisions. This brief discussion has highlighted the major features of the most important types of intellectual property. Engineers seeking more complete information can access the sources provided herein. Important decisions, however, should be made only after consultation with a qualified intellectual property attorney who can tailor an offensive or defensive strategy to suit the real-world environment in which engineers thrive.

REFERENCE

Chisum, Donald S. et al. *Understanding Intellectual Property Law*, Matthew Bender, 1999.